# The History of War Animals

## They also served

Nigel Allsopp

*Dedicated to my granddaughter Alice.*

*You are the angel of my life. You make my heart sing.*

**The History of War Animals: They also served**



ISBN: 978-1-925962-28-4 (paperback)



Cataloguing-in-Publication information for this title is listed with the National Library of Australia.

Published in Australia by **Nigel Allsopp** and InHouse Publishing.
www.inhousepublishing.com.au

Printed in Australia by InHouse Print & Design.

IN HOUSE PTY LTD PUBLISHING

# Contents

# Introduction

During thousands of years of global wars, animals have served alongside man. So, what makes animals useful in war? A dog's qualities of loyalty, intelligence and devotion are highly valued in their role as pets, and these traits are also attractive to the armed forces. Equines such as horses, mules and donkeys have literally carried men on their backs, with all their military supplies to boot. Among their many duties, animal warriors have helped carry messages through the trenches, or flown over them. They have carried ammunition, medical equipment and our wounded. Importantly too, they have given us much comfort and helped to raise morale.

How then, in turn, have we treated them? Sadly, in the past, thousands of animals have been left to unknown fates, on the shores of our enemies, as we sailed home after conflicts.

Today, military working dogs are the predominate animal still used in combat, but marine mammals and equines still hold a position, with the latter having a resurgence in Afghanistan. Coalition Special Forces, including the US, British, and Australian Special Air Services, have used locally sourced horses and mules as transport and pack animals in the mountainous terrain of Afghanistan and the border areas of Pakistan's northwest frontier. One positive side effect has been the training of Special Forces personnel in animal management and basic veterinary skills. While operating in tribal areas, soldiers have won 'hearts and minds' by using this knowledge to treat the sick livestock of local tribes.

# War Pigeons

The first historical mention of a pigeon being used to carry messages in wartime was in the city of Sumer in southern Mesopotamia in 2500 BCE. During WWI, these creatures are credited with helping save the lives of thousands of servicemen, and influencing many key moments in the conflict. Yet, their contribution has been largely overlooked in the intervening decades. They had a success ratio of 95 per cent in getting their messages through, and were particularly useful during battles when field telephones could be disrupted, or once the soldiers had advanced — or retreated — past their prepared lines of communication.

However, as they were trained to fly back to a known base, pigeons could only be used to fly to the rear, rather than to take messages back to the front. The homing ability and navigational skills of carrier pigeons made these birds heroes of both WWI and WWII, and many were awarded for their bravery and heroism. The allies used as many as 200,000 pigeons during WWI, deployed from tanks, aircraft and naval ships.

Pigeons did not have it their own way, of course, apart from soldiers shooting at them from the trenches (a bounty was paid for anyone who killed one), the Germans trained falcons and hawks to attack the pigeons in flight. The reward for the hunter was a good meal.

During WWII, regardless of the arrival of radio technology, pigeons were still used. Carrier pigeons were routinely used by RAF bombers, in an era before GPS and satellite locator beacons. Rescue was far from certain, and a pigeon released as an aircraft was ditching, often brought help.

Australia used 25,000 pigeons during the 'South West Pacific theatre'. When patrols were surrounded by the enemy, or when crews in sinking ships needed rescue, or when medical units were desperately short of blood supplies, the pigeon service came to their aid.

During the D-Day invasion of WWII, when some 156,000 American, British and Canadian forces landed on five beaches along a 50-mile stretch of the heavily fortified coast of France's Normandy region, Allied soldiers maintained radio silence, and relied on a group of 32 pigeons to send messages. Some of these birds also became decorated heroes, awarded with the British Dickin Medal for valour, including famed pigeon William of Orange who, in 1944, when other forms of communication had failed, was released by British soldiers at Arnhem, and flew with a vital message, 400km back to the UK in just under four and a half hours.

As late as 1990, coalition forces in Iraq used pigeons as chemical warfare detectors. They served by dying, to alert troops of a chemical attack.

Even today several armies use pigeons to deliver messages, in case modern electronic systems fail. These include the armed forces of France, Israel and China.

# Veterinary Corps

The British formed the first modern Army Veterinary Service, founded in 1796 by public demand, outraged that more army horses were being lost by ignorance and poor farriery than at the hands of the enemy.

Up until the end of the WWII, horses and other animals were a common sight on the battlefield, either as mounts or as the backbone of the supply train. Many armies in the world developed a veterinary service, in order to ensure that these vital animals were protected from disease and treated for injury. Farriers were an essential part of the welfare system that looked after animals. However, with the advent of mechanised warfare, the majority of service animals were retired, and the supporting veterinary corps and farriers disbanded.

The notion, though, that the military veterinarian became obsolete is not true. Armies around the world have retained or re-established their military veterinary units, which are currently providing valuable service, mainly to military dogs. Today military veterinarians also support humanitarian operations, aiding sick or injured wildlife and farm animals during disaster relief missions. Although, at first, the horse was the main animal to seek care, over time, the military working dog has increased in numbers and is one of the main animals that military veterinarians look after today.

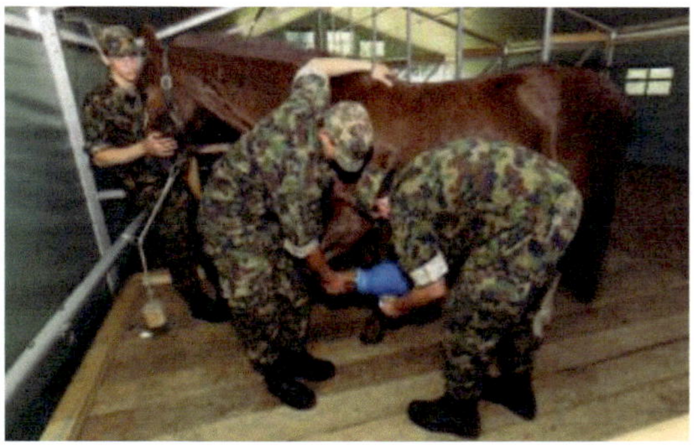

# Mules

In an era when motor transport was still in its infancy and often unreliable, mule power was vital for hauling the guns and supplies needed by vast armies to the frontlines.

No one can argue that the mule is perhaps the most underrated and unappreciated animal that has served in the military. Simply put, without the mule no army in the world, from ancient times to WWII, could have launched any campaign. Perhaps, more mules have given the ultimate sacrifice to man than any other animal. Pack mules provided unlimited mobility to cavalry, infantry, and artillery units. Their downside was that they were fairly noisy and tended to bray during the night, which meant that they easily revealed troop positions to the enemy.

Even though mules were slower than horses, their stamina was much greater than the horse, they could also carry far more weight, and were easier to maintain. Regardless of their reputation for being ill-tempered and stubborn, mules became very popular among the troops.

In WWI, mules carried every supply item you can imagine to the trenches. They often then carried wounded soldiers back to the rear area hospitals too. Mules would go where vehicles could not go. By the end of WWI, the British army owned 213,000 mules.

Brigadier-General Frank Percy Crozier took part in the Battle of the Somme: *"If the times are hard for human beings, on account of the mud and misery which they endure with astounding fortitude, the same may be said of the animals. My heart bleeds for the mules."*

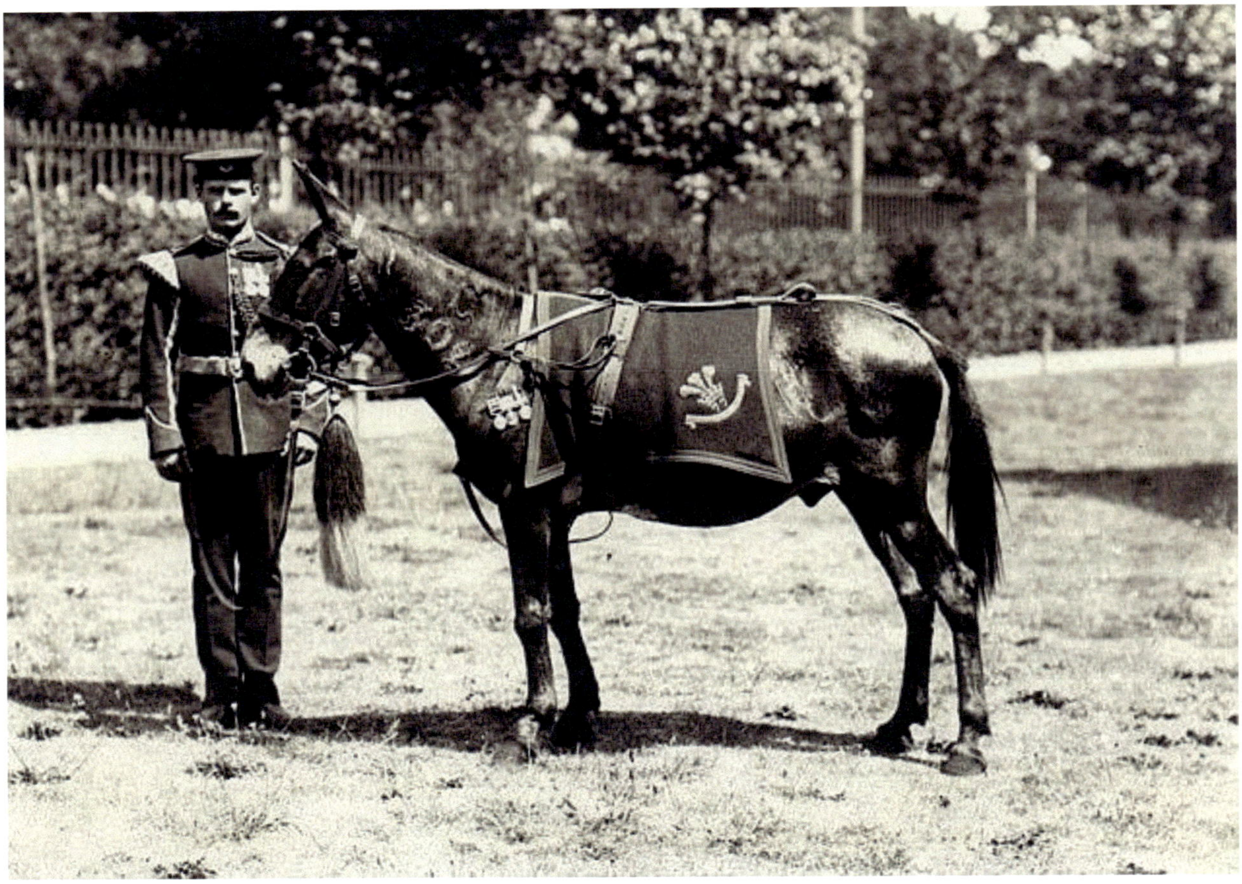

Pictured above stands Jimson, Britain's most highly decorated ass, and the animal with the most medals who served in India and South Africa. He started his military career in India, during the Tirah Campaign between 1897 and 1898, as the British Empire fought the uprising hill tribes in mountain ranges that now form part of Pakistan. Jimson was at the forefront of the expedition, carrying ammunition across difficult terrain. Not only did Jimson have a name when most mules were only given numbers, he was the only mule bestowed with medals for his bravery. After his work with the 2nd Battalion (The Duke of Cambridge's own) Middlesex Regiment in India was complete, and Jimson was taken to South Africa to serve in the Boer War of 1899 to 1902 where his heroics continued. So important was Jimson that the regiment got special permission to bring him back home with them, a very unusual honour for a military animal at that time, due to the risk of disease. After his death in 1914, Jimson lived on, in some way, as the battalion carved an inkwell from one of his hooves.

People think WWII was a mechanised war, but the mule continued to maintain its great value due to its superior ability to negotiate areas inaccessible to tanks and other vehicles, such as the mountains of Italy and the jungles of Burma. A basic guideline used was that a mule could go anywhere a man could go, without the latter using hands. One famous British unit, called the Chindits, used mule columns extensively, some mules carrying heavy machine guns and ammunition, others carrying everything from bandages to beans. They also had the unpleasant ability to be "food on the hoof" for the troops, if needed. After training from the Chindits operating in similar terrain, the US used mules, and this unit became known as Merrill's Marauders.

Chindit mules swam across rivers, climbed thick jungle terrain, or were transported by aircraft, hundreds of kilometres behind enemy lines.

Countries like Brazil, Italy, Switzerland and Germany that used mules to haul light artillery up mountains in the first and second World Wars still use them for the same tasks today.

Today, at the Marine Corps Mountain Warfare Training Centre in California, marines are trained to use mules on combat missions in Afghanistan and other high altitude regions.

# War Dogs

The use of war dogs is not a new concept. The Romans, Egyptians, Greeks, and Babylonians all employed fierce fighting dogs in battle. From these beginnings, dog training and employment has been refined to produce a highly sophisticated and versatile extension of the soldier's own senses. You might think that, with the increased complexity of military operations, the value of war dogs has declined. In fact, the reverse is true. Today's military dog teams are highly deployable and can adapt to many situations on a battlefield. Dogs are ideal for tasks such as tracking, detecting explosives, locating casualties, and guarding facilities.

In WWI dogs were used in a variety of roles. One task was to locate wounded men in "no man's land", and another was to run vital messages, attached to tubes in their collars, from the front lines to headquarters. Large, powerful dogs towed machine guns, and, in the winter snow, they towed sledges with wounded soldiers to aid posts for treatment.

An estimated 100,000 dogs were used by all sides in WWI. By WWII, their numbers had greatly reduced and their roles changed slightly. Alas, some dogs were used on suicide missions against tanks, and others became scouts, walking ahead of patrols to alert to an ambush. Still other army dogs guarded prisoners, or tracked down crashed aircrews behind the front lines.

There have been many brave war dogs. This is Diggers story. Digger was a dark brown and white bulldog who accompanied his owner, Sergeant James Martin, during his service overseas for three and a half years with the Australian army. Digger's remarkable service records state that he went over the top sixteen times, and had survived some of the worst battles of Gallipoli and the Western Front. He had been wounded and gassed at Pozieres in 1916, shot through the jaw (losing three teeth), and was blinded in the right eye and deaf in the left ear. At the sound of a gas alarm, it was reported that Digger would rush to his nearest human companion to have his gas mask fitted. Other accounts note how Digger would take food to wounded men who were stranded in "no man's land". Sergeant Martin returned to Australia in 1918 and was discharged as medically unfit. Digger accompanied him.

One Empire Day (now called Commonwealth Day, an annual celebration of the Commonwealth of nations), Digger was frightened by the celebratory fireworks. Thinking he was under fire again, Digger attempted to jump the fence to warn Martin. Alas, he failed and fell backwards, suffering a burst blood vessel. Digger managed to crawl back into the house. He died on Martin's bed.

*"I often used war dogs in Vietnam in perilous areas where they quite literally saved many lives. There is no doubt that war dogs deserve to be recognized and honoured for their service to our country and the National War Dog Team Monument is a most fitting tribute."*

– General H. Norman Schwarzkopf, USA (Retired)

In Vietnam, dogs were used by Australia and the United States as tracker dogs. They also detected enemy ambushes and saved thousands of soldiers' lives. Many dogs were not allowed to return home, including all eleven Australian dogs.

Today, military working dogs come in all breeds, shapes and sizes, and perform many different roles. Dogs locate explosives and mines, and enhance security by patrolling bases or protecting vital assets or people. These days, a dog may very well parachute into combat with its handler, or be inserted by helicopter or boat into all environments from arctic mountains to desert sands.

*"War dogs have, indeed, served the nation well and saved many lives. Dogs continue to serve to protect both in combat zones and in homeland security roles."*

– General Colin l. Powell, USA (Retired)

Once a military working dog joins a unit, there's little question that humans and non-humans develop emotional ties and a sense of comradeship.

# War Horses

No other animal has played so great a role in the history of warfare as the horse. Archaeologists have found evidence of horses used by raiding nomads as early as 5,000 years ago on the steppes of central Asia and eastern Europe, where it is thought that horses were first domesticated.

Historians estimate that nearly one million horses were used at any one time at the height of WWI. On one single day during fighting, around 7,000 horses died on both the Allied and German sides, due to long-range shelling. During the peak of the war, over 1,000 horses were shipped to Europe from the United States each day.

Lieutenant Dennis Wheatley described the high casualty rate of horses on the front-line in 1915: *"There were dead ones lying all over the place and score of others were floundering and screaming with broken legs, terrible neck wounds or their entrails hanging out. We went back for our pistols and spent the next hour putting the poor, seriously injured brutes out of their misery by shooting them... To do this we had to wade ankle deep through blood... That night we lost over 100 horses."*

Everyone has the image of the dashing cavalry horse, at the charge, saving the day. But horses and ponies of all sizes were used in many ways during both world wars. Small ponies hauled supplies within trench systems, or worked down mine shafts dug by soldiers. The reality is that the vast number of horses had logistical roles, carrying stores and artillery. Most of these work horses were heavy breeds, such as giant Clydesdales and Percherons.

Horses in WWI were deployed to the Western Front and in the deserts of Palestine. One such horse was "Bill the Bastard". Bill was a Waler, which is a hardy type of horse, bred under the extreme climate and challenging working conditions of Australia. Bill was a massive 17.1 hands high, weighed approximately 730 kg and possessed strength and intelligence unmatched by any of his breed. Bill showed courage, endurance and strength that became legendary. He had only one problem — he was the most cantankerous horse that had ever been encountered by the Australian Light Horsemen. It was August 4, 1916. The noise of the battlefield at 2 am was deafening, as the Australians fought the Turks in one of the most important battles, the Battle of Romani. This would mark the turning point of the war in the Middle East. The Australians were severely outnumbered, with a force of 1,700 troopers pitched against 26,000 Turks. When the chips were down, Bill's heroic efforts and exceptional instincts in battle saved the lives of his rider and four other troopers at the Battle of Romani. Bill carried four Tasmanian troops over 3km to safety, away from the Turkish soldiers. Later in the battle of Romani, Bill's master, Major Michael Shanahan, was shot in the leg and passed out. Bill, sensing that his rider was unconscious, carried Shanahan three kilometres to medical aid.

Every country has tales of heroic horses. One million British horses served, yet, thanks to a movie, we are aware of "Warrior", renowned for his amazing courage. His story, like the millions other horses who served, should never be forgotten. When the war came to an end, of the one Million serving horses, 85,000 of the oldest were sold for horse meat, half a million were sold to French farmers to help rebuild the countryside, and only 60,000 made it back to Britain.

Cavalry began to be phased out after WWI in favour of tank warfare, although more horses were used in WWII than in the First World War, particularly by the Axis powers (Germany, Japan and Italy) and Russian armies. The German army entered World War II with 514,000 horses and, over the course of the war, employed 2.75 million horses and mules — the average number of horses in their army has been calculated at 1.1 million. Horse-drawn transportation was most important for Germany, as it was relatively lacking in natural oil resources. By the end of WWII, horses were seldom seen charging across the battlefield, but were still used extensively by many armies for transporting logistics. At the onset of World War II, France mobilised over half a million horses and, as late as 1942, the British army still employed 6,500 horses in the "Mediterranean theatre".

In more recent times, a US marine Mongolian pony called Sgt Reckless showed legendary courage in the Korean War. In 1953, during the Battle of Outpost Vegas, she made 51 unaccompanied trips, carrying 9,000 pounds of ammunition over rugged terrain. She carried wounded marines back, and returned again with ammo. Her handlers taught her to kneel during incoming fire, to step over tripwires, and to memorise routes to and from battle stations.

Reckless was wounded twice and received two Purple Hearts and numerous other awards, including the Dickins Medal for bravery, which is the animal equivalent of the Victoria Cross.

Perhaps one of the world's best-known horses was Sefton, a British Household Cavalry Mounted Regiment horse that survived the Hyde Park bombings, one of the worst terrorist atrocities at the time. The Hyde park bombings killed eleven soldiers and seven horses, and left dozens injured. Sefton survived, despite sustaining 38 injuries, including a partially severed jugular vein and a badly damaged eye. Sefton, who was nineteen at the time of the attack, underwent surgery for more than eight hours — a record for equine surgery. Twenty-eight pieces of shrapnel were removed from his body. He was given a 50 per cent chance of survival, but made a full recovery and returned to duty within three months, serving with the Household Cavalry for another two years. In 1984, aged 21, after seventeen years of military service, Sefton retired to the Horse Trust Sanctuary in Buckinghamshire where he lived until the age of thirty. Today, formal, battle-ready, horse cavalry units have almost disappeared, although horses are still used by organised armed fighters in Third World countries and Special Force teams.

# Donkeys

One of the most famous military donkeys is Simpson's donkey, used at Gallipoli to fetch wounded ANZAC troops down from the front lines to a medical station on the beach. He was just one of many from many nations who took on the same heroic role. During WWI, thousands of donkeys carried loads up to the frontlines on the Western Front, as they were agile and small enough to traverse trenches. The French army, in particular, made great use of them. French donkeys carried 200-pound loads. Tens of thousands were killed, wounded and gassed alongside their handlers. The motor car was still in its early stages of development and was certainly not versatile enough to cross the muddy terrain in France and Belgium, so, in effect, the donkey became the jeep of its day. Many donkeys also served from the heat of the desert campaigns in Egypt to the snow-covered mountains in Italy. 100,000 donkeys were used by Italians alone in WWI. During one campaign in East Africa between 1916-17, 34,000 donkeys supported British troops. Due to deaths caused by tsetse flies, only 1,042 survived the conflict.

WWII saw donkeys used by all sides. Again, their role was similar to that in WWI, and they hauled heavy loads that soldiers could not carry, over terrain that jeeps and trucks could not pass. The Allies used donkeys throughout the Mediterranean, especially during the Italian and Syrian campaigns.

We can still see donkeys in the army today. In recent times, Hughes the donkey helped Canadian Armed Force engineers carry their equipment while on patrol in Afghanistan. Servicemen often had to move big loads through harsh terrain, so the help of the small-but-mighty Hughes was invaluable. Plus, just seeing the cute little donkey put a smile on the soldiers' faces. Life is dangerous for the men and women serving in this region, but animals like Hughes have helped make their load just a little lighter.

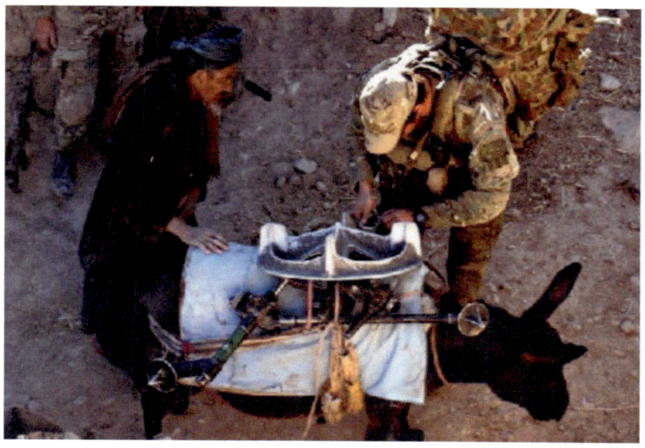

Special Forces in Afghanistan used donkeys to carry equipment over difficult mountainous terrain. They carried laptops and communications gear into remote mountain areas, so that soldiers could call in coordinated precision air strikes on enemy targets, while riding locally purchased mounts.

# Camels

The first recorded use of camels in war was in 853 BCE, when the Arab king Gindibu fielded 1,000 camels in an allied army united against the Assyrians at the Battle of Qarqar, in modern-day Syria. Camels were traditionally preferred to horses in desert settings for their resilience and adaptability, but their unpredictability means they have rarely been used for front-line warfare.

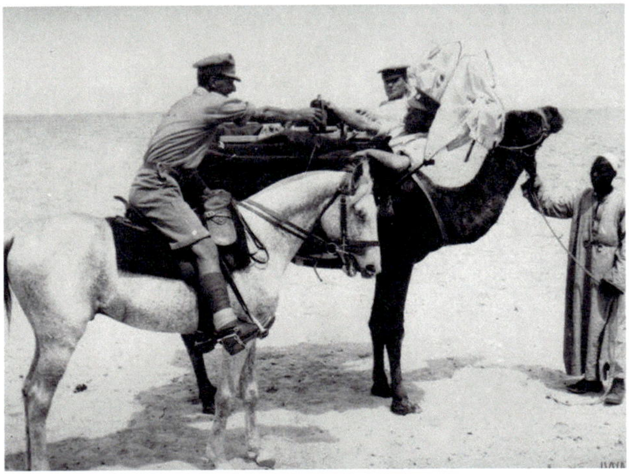

In 1916, the Imperial Camel Corps was formed after the outbreak of a rebellion in the western desert of Egypt. A unit was needed that was capable of traversing large distances over waterless territory. Camels were the obvious form of transport. The men who would serve with them came from England, Australia, New Zealand, India and from across the Commonwealth.

Australia initially sent men from a variety of infantry units who had experience with camels. These men proved successful, as many of them came from Western and South Australian units who had worked with camels in these regions. In late 1917, the Desert Mounted Corps had numbers totalling 6,000 camels. By May 1918, the Corps was reduced in strength to a single battalion, and in 1919 the Corps was formally disbanded. A total of 346 of its personnel were killed in action. Soldiers who served in the Camel Corps learnt to love these animals, and many soldiers were reduced to tears when the Corps was disbanded.

Apart from riding the camels, teams towed heavy artillery and water wagons, plus a special device called a camel ambulance. Most often, camels carried large packs on their backs. Other advantages of camels included that, unlike horses and mules, they did not require shoeing, they could climb mountain trails better than horse-drawn wagons, and they did not get stuck in the mud, like the wagon wheels used by the army at that time. Today the Australian army have two camels as mascots. Several countries and United Nations' forces still use camels operationally to patrol desert terrain. India and several Arab countries also use camels within their border police forces, as they are valued for their ability to endure long marches in harsh, sometimes almost waterless, conditions.

Camels had a bad reputation of being angry and aggressive, but, seen here, is an Australian demonstrating the docility of his camel, placing his wrist in its mouth, in Egypt 1917.

Traditionally, the British Empire has also seen the value of using camels in the military. The US Army too saw their value and imported over 70 camels, used for logistics, between San Antonio and Los Angeles, prior to the civil war. These camels were stationed at Camp Verde, Texas. In 1860, Lt. Col. Robert E. Lee used camels on a long- range patrol. The experiment did not last, however, and the unit was disbanded. Given their proven abilities, camels would have improved logistics in the rugged southwest, during conflicts and garrison resupply operations. The success of camels in French, British, and other armies throughout history appears to validate the US Army camel experiments. Its failure was not due to the camels' lack of capability, rather, the Civil war did not expand westward and so their talents were not needed.

# Mascots

Military mascots have long been of great value to soldiers' morale. From the trenches of WWI to pets adopted by Coalition Forces in Afghanistan today, many troops who missed their own pets, families and loved ones back home have found that an animal in the ranks often gave them a sense of normalcy and friendship, in a place of such horror and destruction.

There are two types of military mascots. Those which appear particularly in Commonwealth forces often have a significant meaning or tradition to the official history with the Regiment. These official mascots are entitled to all the services of the army, including quartering (housing) and food at public expense. Such animals as the Parachute Regiment's Pony, the Royal Fusiliers Blackbuck, The Royal Australian Regiment's 8/9 Battalions Ram, the Irish Guards' Wolf Hound, The US Marines Corps's Bulldog, and Newfoundland of the Canadian Rifle Regiment are a few examples. So, why do we have mascots? The task of caring for a living thing with special requirements helps to enrich the lives of surrounding people. Another reason for mascots is that the presence of a non-human with a formal rank represents a sort of test of protocol and obedience.

The other type of mascot, frequently a dog, is the unofficial pet, which many a soldier has adopted "In-situ" as a companion. This still happens today.

Perhaps one of the most famous mascots during WWI was purchased by Lt. Harry Colebourn of the Canadian Expeditionary Force, on his way to the war in Europe. Colebourn purchased a young bear cub for $20 at a train stop in White River, Ontario. The bear's mother was probably killed in the spring of 1914 when the cub was very young. Colebourn named the bear "Winnipeg", or "Winnie" for short, after his home city of Winnipeg. Winnie accompanied him all the way to England, becoming the mascot of the Canadian Veterinary Corps and a pet to the Second Canadian Infantry Brigade Headquarters. Before leaving for the trenches of France, Colebourn left Winnie at London Zoo.

At the end of the war, instead of taking Winnie back to Canada, Colebourn decided to allow him to remain at the London Zoo where she was much loved for her playfulness and gentleness. Yes, *Winnie the Pooh* was a girl. Among her fans was author A. A. Milne's son, Christopher Robin, who consequently changed the name of his own teddy bear from "Edward Bear" to "Winnie the Pooh", providing the inspiration for his father's story.

Another famed and long-lived connection was between the Welsh Fusiliers and their regimental mascot. From serving with distinction to, more recently, receiving a disciplinary hearing and a demotion for naughty behaviour on parade, the Royal Welch Fusilier goats are called William of Windsor (Billy for short). They have a ration of Guinness, can move up the ranks, are saluted, and wear ceremonial dress when on parade. It all started after a wild goat wandered about the battlefield during the American Revolutionary War in 1775. The goat seemed to lead the Welsh regimental colours to victory at the Siege of Boston.

The first antelope mascot was probably acquired, at some time between 1825 and 1841, by the Royal Warwickshire Regiment, while the unit was serving in India.

Goats too have a long history all over the world as military mascots. This is one breed that has blurred the lines of mascot and military animal. Historically, goats were often carried on navy ships. They were simple to care for, very hardy, would eat just about everything, and had the added benefit of providing milk and meat, thereby preventing scurvy. In WWI, the French army issued milking goats to regiments at the front to provide fresh milk to troops.

In addition to their practical uses, animal companions have often turned into something much more endearing. One such mascot was Smoky, born in Australia but owned by Bill Wynne, an American airman in WWII. Smoky was a tiny Yorkie Terrier, weighing no more than 4 lbs (1.8 kg), but possessing the heart and soul of a lion. Bill and Smoky were inseparable. When Bill got sick and had to return to Brisbane, naturally, Smoky went with him. Doctors at the Brisbane military hospital noticed how well other patients did when Smoky was around. Bill gained permission from the nurses to walk around the wards with Smoky to cheer people up. Smoky would jump on beds and lick the servicemen's faces. Her love was unconditional, and she touched the hearts of many. Smoky's brand of medicine was so successful that doctors recognised a cure for traumatic stress. Smoky became known as the world's first, post-traumatic stress dog. As Smoky's legend spread, she spent half of her tour of duty helping in hospital wards, and the rest of the time deployed to the frontlines with Bill.

Over 75 years passed before Smoky's wartime efforts were finally recognised by the RSPCA who awarded her the Purple Cross Award.

# Animals you might not have thought about

Israel's military has found the perfect vehicle for special operations forces — the llama. After extensive testing, this uncomplaining animal was found to easily outperform donkeys. When the army was training Special Forces to conduct low-signature ground missions in enemy territory, llamas carried more than 50 kg of equipment over mountainous terrain. Israel plans to train a force of llamas to carry up to 100 kg of equipment and supplies. This will ease the burden on troops, and enable special operations' forces to focus on combat or reconnaissance. Donkeys also participated in this exercise at Golan Heights. The donkeys did not perform as well, and required much more food than llamas. Llamas only need to be fed once every two days.

Perhaps one of the most absurd military uses of animals was the US army's idea to strap napalm bombs to bats and send them flying on deadly kamikaze runs. Ultimately known as Project X-Ray, the plan used Mexican free-tailed bats, which were kept calm during travel by forced hibernation. When the soldiers released what would be thousands of napalm-loaded bats, just prior to dawn, they sought dark roost sights in enemy houses. A timer then ignited the bombs.

During the test run, several of the flying bombers unexpectedly took off a little too early and headed, not for their target, but straight for the barracks. They burned down a few barracks together with a General's car. Thankfully, the experiment was stopped.

For hundreds of years, oxen have been used to haul military supplies and artillery. In the Middle Ages, oxen hauled siege engines, during Napoleonic times, they hauled heavy artillery and, in WWII, they took on much the same role in India, Burma and the Chinese theatre of operations. Their relative, the Water Buffalo, also saw action with US troops in the Philippines' war, and the Chindits used them in Burma. Another relative, the Yak, has recently been used in the Chinese army in Xinjiang — high, remote territory on the border with Pakistan, Afghanistan, Kyrgyzstan, Kazakhstan and Mongolia. You can't always count on road access to strategic regions, so soldiers ride yaks.

No, it's not Santa. Reindeer have been used to patrol isolated arctic environments since at least WWI. Still today, the Russian army use reindeer as transportation in the arctic circle, to patrol and protect vital natural oil resources in that region.

In Israel, the armed forces have deployed a herd of antelope as a new frontline force in their battle against Lebanese terror groups. African Eland antelopes too have entered no-man's land between a high security Israeli fences and the southern Lebanese border. This foraging animal performs the vital role of stripping foliage inside the gap that can be used by Hezbollah fighters to stage attacks on Israeli border guards.

# Marine Animals

Most people are familiar with security patrol dogs, and how some service dogs use their keen sense of smell to detect explosives on land. Since 1959, the US navy has trained dolphins and sea lions to help guard against similar threats underwater.

Dolphins naturally possess the most sophisticated sonar known. Mines and other potentially dangerous objects on the ocean floor that are difficult to detect with electronic sonar, especially in coastal shallows or cluttered harbours, are easily found by dolphins. Both dolphins and sea lions have excellent low light vision too, and their underwater directional hearing allows them to detect and track undersea targets, even in dark or murky waters. Unlike human divers, they can also dive hundreds of feet below the surface without risk of "the bends".

During the Gulf War, one sea lion was able to clear the harbour of mines in a few days, allowing ships with vital supplies to use the harbour. Experts estimated that human divers would have taken several weeks to do the same job. Marine mammals are also trained to save lives. They can dive down to a stricken submarine and attach rescue oxygen and communications lines. Someday it may be possible to complete these missions with underwater drones, but for now technology is no match for these animals.

# Small War Animals

Many other creatures, great and small, have served in the military. The ancient Greeks and Romans catapulted beehives over the walls of towns under siege. These stinger-equipped insects were effective weapons when provoked. It's not just due to their sting. Ancient Turkish forces tricked Roman soldiers under the command of Pompey with a tribute to toxic honey. The Romans became drunk and then ill.

This practice continued during the Middle Ages, when attackers continued to launch beehives over castle walls. The same technique was used as recently as World War I, and the Vietnam War Viet Cong guerrillas were said to have carefully relocated wild hives of the Asian giant honeybee, *Apis dorsata,* along the trails used by enemy patrols. One fighter would wait nearby until a patrol approached, before setting off a firework near the hive to aggravate the bees. The bees would then attack the approaching soldiers. Most recently, US scientists have trained bees to search out the odour of land mines.

In 198 BCE, when the Atrenians in the city of Hatra (modern-day Iraq) were besieged, they seem to have perfected a method of handling scorpions without endangering themselves. They filled clay pots with dozens of the stinging creatures and hurled them down on the attacking Roman army.

By the time soldiers noticed the presence of mustard gas on the battlefield, it was often too late. Later, it was discovered that slugs could detect mustard gas well before humans could. These slugs visibly indicated their discomfort by closing their breathing pores and compressing their bodies. Soldiers in the trenches would notice the slugs' response and quickly put on their gas masks to protect themselves from harmful levels of gas. The "slug brigade" saved many lives.

During WWI, using bright lights at night on the frontlines was very dangerous. To avoid being spotted by the enemy, some soldiers instead used glow worms to read maps and important messages after dark. These insects naturally give off a soft, blue-green light. The Australian defence forces used the humble glow worm in the Korean war, smashing them on the rear of their helmets to serve as a night time signal to friendly forces.

# Ceremonial Animals

Our modern-day view of military animals is often via parades and pageants where we view spectacular-looking ceremonial horse riding, ahead of royalty, or on special Veterans' remembrance days. Australia has a long and proud history of mounted troops, both in colonial wars and during WWI in the Egyptian desert. Australia does not have a fulltime ceremonial mounted troop, but several recreation units are scattered throughout the country. They volunteer their time to parade on State occasions and anniversaries.

At Arlington National Cemetery in the US, six solemn horses help deliver departed servicemen and women to their final resting places. These are the caisson horses, who quietly play an important role in the US military, by honouring its service members. A "riderless horse" or "caparisoned horse" is a single horse, without a rider, and with boots reversed in the stirrups, which sometimes accompanies a funeral procession. The horse follows the caisson (a two-wheeled cart designed to carry artillery), carrying the casket. A riderless horse can also be featured in military parades to symbolise fallen soldiers.

Apart from their work during military pageants, a ceremonial horse can call upon the skills of the horseman to perform operational tasks, such as mounted reconnaissance on overseas deployments where no other means of transport works.

The US Air Force at Vandenburgs Base use their mounted horse unit to patrol the vast 100,000 acres and 70 kilometres of coastline of that base. With its difficult terrain, plus a need for conservation as several endangered species reside in the area, horses are the perfect mode of transport.

More recently, ceremonial horses are also being used in therapeutic riding that benefits Veterans with physical injuries. Studies have shown that a horse's gait closely mimics the human stride. When riding, service members exercise the same muscle groups they would if they were walking. Thus, injured soldiers simultaneously train their core muscles, helping to improve their balance and stability. Recent studies show too that therapeutic horseback riding may be a clinically effective intervention for alleviating Post Traumatic Stress symptoms in military veterans. The interaction between horses and riders has been demonstrated to increase riders' confidence, self-esteem, sensory sensitivity and social motivation, while decreasing stress.

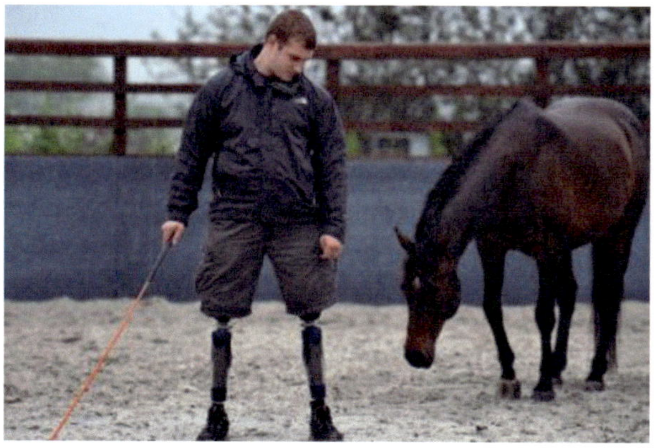

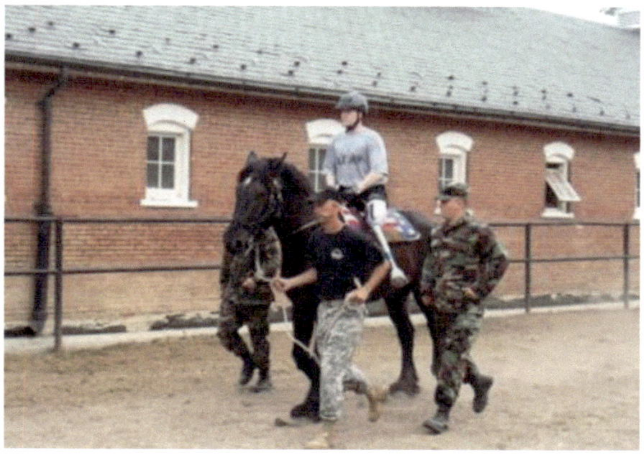

# Elephants

You might be surprised to learn that elephants can still be seen in uniform today! The earliest known military application of elephants, dates from around 1100 BCE in India. The elephant was an important shaper of history too, deciding the fate of many battles in the Greek and Roman world. Beyond raw power, war elephants also brought psychological terror to the battleground.

In the past, the elephant's sheer size caused devastation to enemy troop formations by trampling and impaling soldiers with their tusks. Elephants were sometimes used in siege warfare and for logistical transport, but, most of the time, they acted like a modern-day tank.

The size of an elephant means that it usually needs to be locally sourced, rather than brought any great distance. In WWI, zoo elephants were used to haul heavy equipment on both sides. However, up to the end of WWII, it was with the British army in India where elephants saw most service. In WWII, Field Marshal Sir William Slim spoke of his experience during the Burma campaign, reflecting on these amazing beasts. *"It was the elephants' dignity and intelligence that gained our real respect. To watch an elephant building a bridge, to see the skill with which the great beasts lifted the huge logs, and the accuracy with which they were coaxed into position, was to realise that the elephant was no mere transport animal, but indeed a skilled sapper."*

Elephants next appeared loaded down with rice sacks and ammunition, supporting the Viet Cong in their struggle against US forces. Mud trails in Vietnam would destroy a jeep's chassis, but they were easy to traverse on the back of an elephant. In more recent times, elephants were allegedly used by Iran in 1987 to transport heavy weaponry. Again, as in WWI, these animals were taken from the local zoo.

During the recent earthquake and monsoon flooding emergency, elephants were used for logistical haulage by the defence forces of Myanmar and Thailand to aid civilians. They cleared jungle to get supplies to people, hauled necessary lumber for construction, helped to build bridges, and transported refugees out of the region.

It is ironic that elephants were used recently in Myanmar because, in 1825, the same country is recorded as perhaps the last place where combat elephants went head-to-head with soldiers on the battlefield. They faced the British army and were decimated by firepower. In the modern era, elephants are classed as a pack animal in a US Special Forces field manual, issued as recently as 2004.

# Cats

It's a little-known fact that cats and even kittens accompanied men and women onto the fields of battle, both at sea and on land. "War cats" had two important roles. They were "official mascots" for the regiment, squadron or ship; and unofficial "rat catchers", a less glamorous, though no less essential duty. Cats tend to attach themselves to a territory as much as to people, so where better to find a mascot than in a ship's company? The ship's cat has been a common feature on all naval ships, dating to ancient times. Cats have been carried on ships for many reasons, most importantly to control rodents. Vermin aboard a ship can cause damage to ropes, woodwork, and, in recent times, electrical wiring. One famous naval cat, a black and white feline named Oscar, later became known as Unsinkable Sam. Oscar started his career in the fleet of the German navy, and ended it in the British Royal Navy. He served onboard Bismarck, the HMS Cossack and the HMS Ark Royal, but here comes the cool part — the other thing that all those ships have in common is that all three sank, yet Sam survived them all. Apart from on navy ships, thousands of cats served in WWI trenches. Cats helped the military by killing rats and other vermin that spread disease, and they were sometimes used as gas detectors. Servicemen also found that the presence of cats was good for the morale of men and women who were otherwise engaged in an arguably mundane, yet dangerous environment.

In 525 BCE, according to legend, the Persians used animals, including cats, as shields against the Egyptians during the Battle of Pelusium. The Egyptians regarded certain animals as being sacred, so their religion forbade harm towards them on any account.

# Other Birds

One of the tactics used in WWI was to tunnel under "no man's land" and beneath the enemy's trenches. The tunnels were then filled with explosives and detonated just before an attack. A major risk in mining is suffocating from carbon dioxide or methane, so armies employed the help of canaries. These birds would sing their hearts out while the men worked, but as soon as any gas started filling up the mine, they fell silent. Canaries are very susceptible to a lack of oxygen. The silence of the canaries gave the miners early warning of gas infiltration into the tunnel, allowing them time to escape.

The association between falcons and warfare spans centuries and cultures. Sixteenth century Samurai manuals included a falconry section, and falconry was a component of the education of medieval European knights. As we read in the pigeon's section, when released, pigeons usually brought on a hail of enemy fire, as the Germans tried to bring down the birds and stop their messages getting through. Trained hawks were also brought up to the frontline and sent after the homing pigeons.

The Germans also placed various types of hawks on the coast to hunt British pigeons that were trying to fly to England. The British intelligence service, MI5, began to suspect that the Germans were also employing pigeons, so they established their own "counter pigeon" unit. This was soon abandoned, however, as the hawks from both sides killed pigeons from their own side as well.

In modern times, this process is still used by the Belgian Air Force who have a specialist unit of birds of prey, trained to kill and ward off flocks of birds from military airfields, which might otherwise cause aircraft crashes.

# War Animals in the Future

Since man has gone to war, he has felt the need for animal companionship. Soldiers in many historical photos can be seen with pet cats, birds and dogs, and an assortment of agricultural animals, like chickens, ducks and sheep. Soldiers have even taken their own country's native animals, such as kangaroos or kiwis to war, or picked up and befriended local, exotic animals, such as monkeys.

Most would agree that, if possible, we should stop sending our pets to war. So, why are these animals still necessary? The simple answer is that no technology exists that can replace their abilities to save man's life during war. For example, a dog can detect the scent of a road side bomb many meters away, and so save soldiers' lives. The number of military dogs throughout the world is on the increase, and many more are used today than during WWI. Canines are not just helping soldiers on the frontlines, but also on their return home because various governments recognise the benefits of specifically trained dogs as part of therapeutic treatment, for both physical and mental health conditions, including post-traumatic stress disorder (PTSD). Dog companionship has reduced symptoms resulting from trauma that are often triggered by repeated exposure to combat situations.

With today's immigration and border control concerns, some countries have rediscovered the use of horses to patrol difficult terrain, such as in the high, alpine terrain of Europe. Horses provide fast mobility and a great observation platform. They are silent and can outrun illegal immigrants.

As drones become more affordable, and increasingly threaten to fly into security-sensitive zones, several air forces around the world have sought anti-drone technology. The French air force has showcased its latest anti-drone weapon — an eagle, named D'Artagnan. Golden Eagles combat these battery-charged drones by aerial attack, bringing them to the ground.

People might think that animals would not be much help to soldiers in today's high- technology world, but sometimes the old ways are still the best! Modern technology has also been used to try and replace the mule and donkey, which can carry excessive loads over difficult terrain where vehicles cannot go. After years of experiments and millions of research dollars and prototypes, the military have returned to the reliable mule.

*"The capability Military Working Dogs bring to the fight cannot be replicated by man or machine. By all measures of performance, their yield outperforms any asset we have in our inventory. Our army [and military] would be remiss if we failed to invest more in this incredibly valuable resource."*

– General David H. Petraeus, USA

*"Technology, like the electronic frequency jammers used in Iraq to prevent bombs from being detonated by cell phones or TV remotes, has been successful, but the highest detection rates were still achieved using K-9 units and trained handlers."*

– Lt. General Michael l. Oates, USA

# Memorials

Human and animal deaths are a tragic consequence of war. It is only fitting that the animals pressed into service be honoured for their loyalty and courage for the part they have played in history.

The most important WWI Australian war animal memorial is in Pozieres, France. Established in 2017, it lies in the heart of many battlefields of WWI and honours war animals from all nations and all sides. The memorial, called *Emerging Spirit*, displays a mule, a donkey, a dog and a pigeon within the horse's mane.

The New Zealand war animal memorial, at Waiouru National Army Museum, honours all NZ War animals in all wars. The millions of horses, mules and donkeys from all sides who never returned from the First World War have been honoured with their own memorial in Britain. Named "Poppy the Warhorse", the monument was unveiled in June 2018 at a ceremony in Ascot — on a site looking across to one of the country's best-known racecourses. The larger than life bronze figure, standing on a three-metre high stone plinth was described in a message of support from Queen who wrote: *"It is my sincere hope that the memorial provides an opportunity to reflect on the sacrifice made by animals during the Great War."* The Australian War Animal Memorial Organisation has established several memorials around the world. It continues to educate and remind people of their deeds and sacrifices.

# A Note from the Author

In this book, we have explored a brief history of military animals throughout the ages, their different roles, what animal warriors do today, and what the future holds for animals in modern warfare. One wish of mine in writing this book is that, on future Memorial Days, when we think of all our soldiers who gave their lives in war, that we also pause a thought for the four-legged heroes that fought and died alongside them. Hopefully, one day, we can not only stop sending our pets to war, but our sons too.

> *"The greatness of a nation and its moral progress can be judged by the way its animals are treated."*
>
> – Mahatma Gandhi

# Animals of War
# We Will Remember Them

Their blood stained the land, as they served with pride.
A duty to their master, they stood side by side.
From the sky to the sea, through a vast open land,
Together they fought, both animal and man.
They flew through the sky, as messengers on wing.
Dodging bombs and bullets, for salvation to bring.
Their wings did strain, till their hearts near burst,
They flew day and night, never fearing the worst.
They walked together midst a bomb-spangled field,
Both dog and handler, neither one would yield.
The bond and the friendships they shared through war,
Will live and be remembered for ever more.
Through the sand so fine, they lolloped to the fore
They're the ships of the desert, the Camel Corp.
Over hills and through valleys, the line did twine
As they carried their loads to the firing line.
Into combat they rode, both man and steed,
Through the fear of battle, they were a special breed.
The mateship of horse and rider was strong
Made the parting in death, seem so heartless and wrong.
As beasts of burden, they were put to the test,
With hearts filled with anguish, they gave their best.
They died where they fell, while the others pulled on.
Are they lost in life's story? Can we still hear their song?
The theatre of war — hell for animal and man,
Whatever the cost, they cannot understand.
So, remember them kindly, as you walk through life,
For they too served proudly, to help make things right.

– *Santina Lizzio*

Made in the USA
Las Vegas, NV
07 November 2026

57669930R00026